From Animal Kingdom to Human Mind: Tracing the Origins of Instinct and Nature

---

Abdul Basit

About the Author:

Abdul Basit is a renowned philosophical author with a profound expertise in modern metaphysics and divine philosophy. He obtained his postgraduate degree in Philosophy from the prestigious University of the Punjab, where he specialized in modern metaphysics during the period of 2007 to 2009.

Throughout his academic journey, Abdul Basit demonstrated an exceptional aptitude for grasping the intricate concepts of modern metaphysics, making it his favorite subject of study. His deep perceptual knowledge of this field allowed him to delve into the depths of philosophical inquiry, enabling him to explore and write extensively on various metaphysical topics.

With a dedication to his craft, Abdul Basit has authored numerous articles on a wide range of philosophical subjects, emphasizing his expertise in metaphysics. His writings exhibit his remarkable ability to articulate complex philosophical concepts in a concise and accessible manner, making his works engaging to both scholars and enthusiasts alike.

Having dedicated more than two decades to research and studies in metaphysics, Abdul Basit's findings have led him to the conclusion that while human intellect may be fallible, divine knowledge remains absolute. His research has particularly focused on various divine books, leading him to a personal realization that the original divine knowledge exists within the Holy Quran.

Abdul Basit's commitment to pursuing intellectual excellence, combined with his deep understanding of modern metaphysics and divine philosophy, has established him as a leading authority in the field. Through his writings, he continues to inspire readers to explore the profound mysteries of existence and contemplate the intricate interplay between the human intellect and divine knowledge.

With an unparalleled passion for philosophy and a relentless pursuit of truth, Abdul Basit's works serve as a beacon of enlightenment, inviting readers to embark on a journey of profound self-reflection and intellectual growth.

CONTENTS

PREFACE

In the vast realm of knowledge, the enigmatic workings of the animal kingdom and the intricate depths of the human mind have always captivated the curiosity of both scholars and laypeople alike. As we navigate the boundless expanse of the natural world, we are confronted with a fascinating dichotomy—Nature and Instinct. These two pillars, seemingly intertwined, shape the very essence of our existence and hold the key to understanding the intricate tapestry of life.

In this short book, "From Animal Kingdom to Human Mind: Tracing the Origins of Instinct and Nature," we embark on a journey of exploration and revelation, delving into the depths of these fundamental concepts that define our world. With a keen focus on deciphering their origins and unraveling their intricate connections, we seek to shed light on the fundamental differences that set Nature and Instinct apart.

Drawing upon a myriad of modern references and scientific discoveries, we delve into the rich tapestry of knowledge accumulated by researchers and scholars who have dedicated their lives to unraveling the mysteries of our existence. Through their pioneering work, we uncover the underlying mechanisms that govern the animal kingdom's innate behaviors and the intricate complexities of the human mind.

However, this exploration transcends the confines of scientific inquiry. We strive to bridge the gap between science and spirituality, recognizing that our quest for understanding cannot be complete without acknowledging the existence of something greater. In our pursuit, we turn to the timeless wisdom contained within a divine book, seeking the profound insights it offers on the origins of Nature and Instinct.

It is our hope that this short book will prove to be a valuable resource for students researching these compelling topics, as well as for researchers around the world who seek to deepen their understanding of the human mind and the animal kingdom. We aspire to provide a fresh perspective, one that challenges conventional wisdom and offers a unique blend of scientific exploration and spiritual contemplation.

As the pages unfold, we invite you, dear reader, to join us on this intellectual odyssey. Prepare to be enthralled by the captivating interplay between Nature and Instinct, and to embark on a thought-provoking exploration that will expand your horizons and ignite your imagination. We aim to present you with a wealth of knowledge that will both stimulate your mind and leave you with a deep sense of wonder and enlightenment.

May this book serve as a guiding light, illuminating the path towards a more profound understanding of our place in the grand tapestry of existence. We sincerely hope that it will inspire you to question, to explore, and to delve into the remarkable depths of the human mind and the captivating wonders of the animal kingdom.

With heartfelt anticipation, we embark on this journey together, confident that the insights you will glean from these pages will forever transform your perception of Nature and Instinct. Prepare to unlock a world of knowledge and embark on a voyage of discovery that will forever alter your understanding of the innate forces that shape our world.

Welcome to "From Animal Kingdom to Human Mind: Tracing the Origins of Instinct and Nature."

Happy reading!

Abdul Basit

Chapter No.1

Introduction

وَمَا تَوْفِيْقِيْ اِلَّا بِاللّٰهِ عَلَيْهِ تَوَكَّلْتُ وَاِلَيْهِ اُنِيْبُ

And my success is not through Allah. Upon him, I have relied, and to Him,
I return.
Chapter No. 11 Ayat NO. 88

Nature began at the same time when Allah created this universe from nothingness. This universe operates by the laws of nature. As far as we can observe the universe, it will appear to us to be bound by the laws of nature. The matter of the universe flows around us in the infinite void according to the laws of nature. The orderly way in which the entire universe operates is solely due to the laws of pure nature.

One thing is clear from the above, the matter of the universe is bound by the laws of nature.

Allah (God) has created innumerable living beings from this inanimate matter. The body of living beings is material and the matter is subject to the laws of nature. There is another thing in living beings which is called instinct. Instinct is found only in living beings. Inanimate matter has no instinct. The movements we see in living beings are motivated by instinct. Living beings move from one place to another due to instinct.

There are two types of movement in living things, one movement is due to instinct and the other movement is due to the laws of nature. For example, when there is a strong wind, the direction of the flying birds is seen to change due to the pressure of this air. If when it rains, the birds cannot fly because of the rain, it is also due to the laws of nature. Two things are clear from this conversation: there is a difference between instinct and nature. And we come up with two principles.

- **Matter is controlled by the laws of nature.**
- **Instincts exist only in living beings and because of instincts living beings perform movements.**

Keeping the above principle in mind, let's see what the experts and thinkers of the world have explained about instinct and nature. First of all, I will explain the meaning and definition of instinct, but before that, it is important to understand the difference that among all living beings, human beings are also included, and human beings are the only rational animals that have intellect and consciousness in addition to instinct. Other organisms do not.

Keeping this fact in mind, I explain the definition of instinct in English which meets the above principle.

Definition of Intuition, instinct and habit

"An intuitive reaction not based on rational conscious thought."

According to this definition, instinct is a reaction in an organism that does not involve conscious action. Instinct is a trait further described as:

Instinctive behavior

Instinct is the ability of an animal to perform a particular behavior in response to a given stimulus the first time the animal is exposed to the stimulus. In other words, instinctive behavior does not have to be learned or practiced.

In animal biology

Jean Henri Fabre (1823-1915), an entomologist, considered instinct to be any behavior that did not require cognition or consciousness to perform

In psychology

The term "*instinct*" in psychology was first used in the 1870s by Wilhelm Wundt. By the close of the 19th century, most repeated behavior was considered instinctual

Instinct has nothing to do with learning and no living being can acquire it through any special process, rather this ability is naturally present in living beings by nature. For example, feeling hungry or thirsty. When an organism feels hunger or thirst, the instinct within it compels it to satisfy

hunger and thirst and this action is due to instinct. Different thinkers of the world have defined instinct differently. Now we read its description which is as follows.

Definitions of 'instinct

1. a fixed action pattern, in which a very short to medium length sequence of actions, without variation, are carried out in response to a clearly defined stimulus. *"Karl Lorenz"*
2. a propensity before experience, and independent of instructions. *"William Paley"*
3. a blind tendency to some mode of action, independent of any consideration, on the part of the agent, of the end to which the action leads. *"Whately"*

4.	an agent who performs blindly and ignorantly a work of intelligence and knowledge.

	"Sir W.Hamilton"

5.	The resemblance between what originally was a habit, and an instinct becomes so close as not to be distinguished.

	"Charles Darwin"

6.	a natural, unreasoning, impulse by which an animal is guided to the performance of any action, without improvement in the method.

The explanation of the above-mentioned different thinkers, the meaning of instinct comes to us that instinct is only a characteristic in living beings due to which any living being continues its journey of life. Another explanation is necessary here that instinct has another thing called habit. Habit can also be developed in living things, but habit is not created in living things, but repeating a specific action of any living thing over and over again and repeating that action until it becomes habituated to that action is called a habit is very different from instinct and nature.

Habit

"Routine of behavior that is repeated regularly and tends to occur subconsciously".

The American Journal of Psychology (1903) defines a *"habit*, from the standpoint of psychology, as a more or less fixed way of thinking, willing, or feeling acquired through previous repetition of a mental.

From the explanation given above, it is known that habit has nothing to do with instinct because instinct is instilled at birth whereas habit later on by repeating a certain action over and over again the organism becomes accustomed to that action and so it would become a habit of it.

Reality of Nature?

After instinct, we now understand nature, what is nature? Different thinkers have explained it differently. Considering the principle of nature above, let's see what the thinkers of the world think about it. According to the principle given above, we know that the matter in the universe is running according to the laws of nature and all the changes taking place in it are happening according to the laws of nature. Let's first see how it is defined in the English language.

The word *Nature is borrowed from the Old French nature and is derived from the Latin word nature, or "essential qualities, innate disposition", and in ancient times, literally meant "birth". In ancient philosophy, nature is mostly used as the Latin translation of the Greek word physis (φύσις), which originally related to the intrinsic characteristics that plants, animals, and other features of the world develop of their own accord.

In the above, the meaning and explanation of the word Nature have been explained in its language and history. This word is considered synonymous with the Arabic and Urdu word Fitrat. But the Arabic language has many terms which we make synonymous with other languages and due to this, there is a difference in meaning. Now we see what is the Qur'anic meaning of nature, which is mentioned in Tabveeb-ul-Quran (written by Ghulam Ahmad Pervez,) like this.

"*Nature*"

(Laws of Nature. Phenomena of Nature)

Nature (ف ط ر) root word

Thus, the basic meaning of this substance is to divide. to tear digging or building but the basic condition is that this work should be done first. Therefore, the meaning of Fitrat (nature) will be that he did this work for the first time. For this reason, God is said to be the creator of the heavens and the earth (6:14). That is, He created the universe for the first time. He

who brought it into existence from nothingness. The same has been called badi' al-samwatwal-arz

بَدِيْعُ السَّمٰوٰتِ وَالْأَرْضِ

in another place. (6:102). By this Nahj, Fitrat(nature) Allah will mean the law of God's creation

according to which He brought the things of the universe (and man) into being from nothingness. From which He created them.

In the above explanation we have seen how the Qur'an describes nature. The laws of nature are irrevocable and cannot be compromised. Among living things we now come to man. In addition to humans, other insects and animals have instincts, but humans also have intellect and consciousness in addition to instincts, due to which humans are called Ashraf al-Makhluq (superior creature). Let us see how philosophy explains human consciousness as distinct from instinct.

Chapter No.3

Divine Introduction of man on earth

Man is a rational animal, as "Lecomte du Novy" says. What properly defines and distinguishes a human being as a human being is existence within him of abstract concepts of moral ideas and spiritual feelings. There are other qualities that a person can rightly be proud of. These concepts are as real as the body itself and those that make the figure possess importance and value -- otherwise, without these things, the existence of the figure is just a figure.

The scientific study of the anthropological element has highlighted the facts about human beings. As an animal and from the point of view of physiology, man is a better animal than all other species. The human nervous system is superior to that of other animals. Stimulated Response and Conditioned Response, this character formula is also unable to explain all the human factors. In the process of human perception, there is a strong tendency to dispose of symbols so that they can be given meaning. It is not just a simple emotional reaction, but it is fundamentally a process of interpretation and understanding. Human knowledge is not always related to the process of "trial and error" but more often it is related to thinking,

As well as the disposal of symbols and insights. The most important part of human knowledge is deductive. Man, moving from the known towards the unknown and from the experimental to the non-experimental, forms a higher body of his knowledge based on his specific abilities. He makes new plans from his actions and moves from one triumph to another - and one who meditates, learns, remembers, plans, imagines, and derives the absolute facts of his experience and all these factors are primarily intellectual, and implicitly sense a dynamic nature. Therefore, the intellectual factors indicate the existence of some kind of soul and self.

In light of the above philosophy, we have come to know that there is something that exists in man that separates man from other creatures. But there is another fundamental difference between man and other living beings, and that difference is that the guidance for living in this world is already present in all living beings except man, and each living being has a different There is guidance in form. For example, in the bee, this guidance has already been deposited to make honey in the beehive by sucking the sap from the gardens. And so the other insects and animals have their distinct abilities naturally present within them. There is also a description of the bee in the Qur'an, which means something like this.

وَأَوْحٰى رَبُّكَ إِلَى النَّحْلِ أَنِ اتَّخِذِيْ مِنَ الْجِبَالِ بُيُوْتًا وَّمِنَ الشَّجَرِ وَمِمَّا يَعْرِشُوْنَ

(سورة النحل , آيت 68)

(If you want to see how the law of God guides in the universe and how everything performs the actions of the wise under His guidance) then look at the bee. He has placed the guidance in it that he should make his beehive in the mountains, in the trees, and in the beehive that is made for this purpose.

chapter No.16 verse 68

Apart from this, other animals are also mentioned in the Holy Quran which is mentioned below

Mention of animals in the Qur'an

Two hundred verses of the Qur'an are about animals, and a total of thirty-five animals are mentioned by name in the Qur'an. They include birds, insects, wild and domesticated animals, etc.

As far as man is concerned, there is no pre-existing guidance within man, but man's mind is like a blank paper on which man writes by learning from his environment.
There is external (divine) guidance for man too, by following which constructive abilities can be created within man, but since man is born free and man has also been given willpower, a man in his egoism goes to the level due to which evils and inequalities arise in human society. The Holy Qur'an has also described the psychology of man that when a man puts aside the external(divine) standard, then what is the condition of a man? Let's know from the holy Quran.

Chapter No.4

How the Divine Book Quran Explains the Psychology of Humans?

What the Holy Qur'an itself has said about man also makes it clear that these can never be demonstrations of "God's nature". (It should be clear that the Holy Quran has said something about the Man who does not follow the guidance of revelation but follows his feelings). For example•

1.   اِنَّ الْاِنْسَانَ خُلِقَ هَلُوْعًا

Man is very impatient. His intention is not fulfilled.

Chapter No.70 Verse No.19

2.   {اِنَّهُ كَانَ ظَلُوْمًا جَهُوْلًا}

Very cruel and ignorant

Chapter No.33 Verse No.72

3. قُتِلَ الْإِنْسَانُ مَآ أَكْفَرَهُ

Very ungrateful.

Chapter No.80 Verse No.17

4. وَكَانَ الْإِنْسَانُ عَجُولًا

Very hasty

Chapter No.17 Verse No.11

5. وَكَانَ الْإِنْسَانُ أَكْثَرَ شَيْءٍ جَدَلًا

There is often a dispute about things.

Chapter No.18 Verse No.54

6. فَإِذَا هُوَ خَصِيْمٌ مُبِيْنٌ

Very quarrelsome.

Chapter No.36 Verse No.77

Above, you have seen that when a person is not bound by external(divine) principles, what kind of emotions arise in his psyche?

Man needs guidance which can be found only in the Holy Qur'an. There is no doubt that whenever man ignores permanent values, there will be inequalities in society. And there is no doubt that man has no nature. If man had nature, man would be free. It would not have happened, but the laws of nature would have bound his inner being, but human beings are free in their inner state. Now we see what is the meaning of permanent values in the Holy Quran.

Permanent values

The Holy Qur'an has given some commands (or laws) to guide mankind and the rest has given principles. These principles are also called Permanent values and since these principles are immutable, they are permanent values. (PERMANENT VALUES) is defined as the term.

There are many principles in the Holy Qur'an of permanent values, but I list a few principles(permanent values) here.

Permanent Values and Principles of Life

1.   *Respect humanity*
2.   *Determined levels*
3.   *The universe is created in truth.*
4.   *Creation process*
5.   *the blessings of life*
6.   *Knowledge and wisdom*
7.   *Emotions (within Quranic limits)*
8.   *freedom (within Quranic limits)*
9.   *No compulsion in religion(islam)*
10.   *Justice*
11.   *Unity of Ummah*
12.   *Unity of humanity*
13.   *Balance in values*

The above-mentioned principles are irrevocable, there cannot be any deficiency in them, and time is a witness that whenever a nation violated these principles, the irrevocable law of retribution destroyed that nation.

We are talking about instinct and nature and the purpose of all this explanation is to give us the correct meaning of nature and instinct. Because we have some wrong concepts of nature. One concept is that human beings also have nature, but the purpose of the explanation above is to know the difference between instinct and nature. And it has been proved that man has no nature, but the material body of man is governed by the laws of nature.

And we know that (God) Allah created living things from inanimate matter. The physical existence of an organism starts from the (smallest tiny) point and this existence increases with time and ends up at a certain volume, which is called the death of this organism. Allah has created every living thing in the world with different forms and abilities, And according to modern research, the unit of the body of all living organisms is a cell. Tissues together make up organs. Organs together make up systems and systems make up the physical being of any organism. And so there is a complex system inside the cell. This cell has a nucleus and inside.This nucleus is the DNA system. According to Wikipedia, Let's read modern research on DNA.

Chapter No.5

DNA and Birds migration

What is DNA?

Deoxyribonucleic acid (/diːˈɒksɪ ˌraɪboʊnjuː ˌkliːɪk, - ˌkleɪ-/ (listen); DNA) is a polymer composed of two polynucleotide chains that coil around each other to form a double helix. The polymer carries genetic instructions for the development, functioning, growth, and reproduction of all known organisms and many viruses
(Wikipedia)

It is just as the HTML codes are behind the page visible on the computer browser, DNA codes are behind the moving life on earth. That is, the appearance and behavior of an organism (phenotype) are formed by the genetic codes hidden in the DNA in its cells. / is called genotype. The difference between style and style is explained as if a drama seen on a TV screen runs entirely on a script written for itself as if the drama itself is an example of style and the script written for it is style.

From the above information, we have come to know that Allah(God) has placed the ability of different systems in every living being. Which develops gradually and when the organism reaches its peak, we will see his ability in the environment. For example, animals have their abilities and birds have their abilities. Some birds migrate from one place to another which is described as follows

Bird Migration

Bird migration The sequential movement of birds due to changes in weather can be called bird migration. And this migration is done only once a year. Thus a flock of birds migrates from one place to another suitable place.

Apart from this, I am presenting you the details of ten famous migrations of living beings in the world. By reading this, the intellect is stunned as to how beautifully (God) Allah has created every living being. The description of the migratory creatures is as follows

Top Ten Most Amazing Migrations

1. *Arctic Tern (Sterna paradisaea)**
   *Travel**: 71,000 kilometers a year.
   *From**: Greenland and the Arctic to Antarctica.

2. *Humpback Whale (Megaptera novaeangliae)**

   *Travel**: One female humpback whale traveled more than 9,800 kilometers.
   *From**: move from the tropics and head north to feeding grounds. Not all travel together; pregnant whales and those who had calves in the previous year go north first.

3. *Sooty Shearwaters (Puffinus griseus)**

   *Travel**: 65,000 kilometers
   *From**: travel from breeding grounds in New Zealand and Chile north to feeding grounds covering around 724 to 1096 kilometers a day.

4. *Monarch Butterfly (Danaus plexippus)**

Travel: 3,100 kilometers
From: arrive in Canada in June, then in September (two to three
enerations later) head south to Mexico.

5.   *Dragonflies mainly the Globe skimmers (Pantala flavescens)*

Travel: 14,000 – 18,000 kilometers
From: he heads out from India to the Maldives, Seychelles,
Mozambique, and Uganda, using the wind to help him along. go through 4
enerations for the complete migration cycle.

6.   *Chinook Salmon (Oncorhynchus tshawytscha)*

Travel: 3,000 kilometers
From: After hatching he spends time in freshwater from three months to
 year. He migrates to the Pacific Ocean, then heads back home to the river
e was born in to spawn.

7.   *Adélie Penguins (Pygoscelis adeliae)*

Travel: 17,600 kilometers
From: He follows the sun from the breeding colonies to winter feeding
rounds.

8.   *Semipalmated Sandpiper (Calidris pusilla)*

Travel: 3,000 kilometers
From: In mid-May, he takes off from South America heading north

towards his breeding grounds in the sub-arctic of Canada and Alaska. **In July, He head back south again.**

9.   **Wildebeest or Gnu (Connochaetes)**
        Travel: The Serengeti population of wildebeest is a huge nomadic group that migrates 1,600 kilometers each year.

From: Beginning in January and February we move from the Serengeti plains west towards Lake Victoria.

10.   **Red Crab of Christmas Island (Gecarcoidea natalis)**

Travel: 5 kilometers, traveling up to 12 hours over 5 days.

From: At the beginning of the wet season (October/November) He heads out from the forest to the coast to breed. The males arrive at the sea first followed by the females who soon outnumber them. As tiny babies (only 5 mm across) They travel back from the sea to the forest, a trip that takes about nine days.
(Wikipedia)

From the above information, one thing becomes clear in all living things other than man, some guidance has been given by nature, but man has no nature, man is a product of his environment.

Chapter No.6

Conclusion

Finding of my discussion

Today's topic "Distinctions in Instinct and Nature" has been discussed and there is a significant difference between instinct and nature. That man has no nature. The material existence of man and other living beings is governed by the laws of nature. Apart from human beings, only instinct is found and every living thing has the common instinct of hunger, thirst, and reproduction, but other than that, the other movements and habitats of living beings are different, which have been entrusted to them by nature. In addition to instinct, man also has intellect and consciousness, due to which he is called Ashraf al-Mukhluq اشرف المخلوق (superior from other species) among other living beings. Man has no nature by birth, but the material body of man is bound by the laws of nature. Man has a willpower that overrides human instinct. For example, when a man feels hunger and thirst, man can control it with willpower, but other animals do not have willpower, so they start looking for food. In short, man has no nature and his mind is clean like a blank paper and man learns from his environment and writes on the blank paper of his mind.

conclusion

In conclusion, I want to say that the truth of instinct and nature has a small hope of effort that I hope the reader will have insight from. May Allah (God) make us all aware of the truth and may Allah (God) be our supporter and helper. Amen!

رَبَّنَا تَقَبَّلْ مِنَّا إِنَّكَ أَنْتَ السَّمِيعُ الْعَلِيمُ، وَتُبْ عَلَيْنَا إِنَّكَ أَنْتَ التَّوَّابُ الرَّحِيمُ

Notes and References

1.   *Introduction to Philosophy of Man*
      "Fundamental Problems of Philosophy"
      Urdu book of " Qazi Qaiser-ul-Islam"

2.   *English meaning and definition of instinct*
      "Wikipedia"

3.   *English meaning and definition of habit*
      "Wikipedia"

4.   *English meaning of nature*
      "Wikipedia"

Printed in Great Britain
by Amazon

25970168R00020